United States Nuclear Regulatory Commission

Protecting People and the Environment

NUREG-2154

I0493855

Acceptability of Corrective Action Programs for Fuel Cycle Facilities

Draft Report for Comment

Office of Nuclear Material Safety and Safeguards

AVAILABILITY OF REFERENCE MATERIALS
IN NRC PUBLICATIONS

United States Nuclear Regulatory Commission

Protecting People and the Environment

NUREG-2154

Acceptability of Corrective Action Programs for Fuel Cycle Facilities

Draft Report for Comment

Manuscript Completed: January 2013
Date Published: January 2013

Prepared by:
Sabrina Atack

Office of Nuclear Material Safety and Safeguards

COMMENTS ON DRAFT REPORT

Any interested party may submit comments on this report for consideration by the NRC staff. Comments may be accompanied by additional relevant information or supporting data. Please specify the report number NUREG-2154, in your comments, and send them by the end comment period specified in the *Federal Register* notice announcing the availability of this report to the following address:

Cindy Bladey, Chief
Rules, Announcements, and Directives Branch
Division of Administrative Services
Office of Administration
Mail Stop: TWB-05-B01M
U.S. Nuclear Regulatory Commission
Washington, DC 20555-0001

For any questions about the material in this report, please contact:

Sabrina Atack
Mail Stop E2 C40M
U.S. Nuclear Regulatory Commission
Washington, DC 20555-0001
Phone: 301-492-3204
E-mail: Sabrina.Atack@nrc.gov

Please be aware that any comments that you submit to the NRC will be considered a public record and entered into the Agencywide Documents Access and Management System (ADAMS). Do not provide information you would not want to be publicly available.

ABSTRACT

The NRC staff has revised Section 2.3.2 of the NRC Enforcement Policy (ADAMS Accession No. ML12340A295) to disposition Severity Level IV violations for Fuel Cycle Facilities as non-cited violations if the NRC determines that the licensee's CAP is effective, the licensee enters the violation in its CAP, and other criteria in Section 2.3.2 of the Enforcement Policy are met. The purpose of this draft NUREG, "Acceptability of Corrective Action Programs for Fuel Cycle Facilities," is to provide guidance to the NRC staff on how to determine, from a licensee's CAP licensing submittal, that a CAP is acceptable. After the NRC staff determines that the CAP is acceptable, the CAP licensing submittal will be incorporated into the license and implementation of the CAP will be verified by an NRC inspection using a CAP inspection procedure. After the NRC inspection verifies that the licensee has implemented its CAP in accordance with the license and the licensee's CAP implementing procedures, then the NRC will consider the CAP to be effective for the purposes of Section 2.3.2 of the Enforcement Policy.

CONTENTS

INTRODUCTION

In the staff requirements memorandum (SRM) for SECY-10-0031, "Revising the Fuel Cycle Oversight Process," dated August 4, 2010 (Agencywide Documents Access and Management System (ADAMS) Accession No. ML102170054), the Commission directed the U.S. Nuclear Regulatory Commission (NRC) staff to consider how the NRC Enforcement Policy could best reflect that most fuel cycle licensees have voluntarily developed corrective action programs (CAPs). In response to the Commission's direction, the staff revised the NRC Enforcement Policy to disposition Severity Level IV violations as noncited violations if the NRC determines that the licensee's CAP is effective, the licensee enters the violation in its CAP, and other criteria are met, as delineated in Section 2.3.2 of the NRC Enforcement Policy (ADAMS Accession No. ML12340A295). In SRM-SECY-11-0140, "Enhancements to the Fuel Cycle Oversight Process," dated January 5, 2012 (ADAMS Accession No. ML120050322), the Commission directed staff to proceed with the development and implementation of the incentives for licensees to maintain an effective CAP.

All licensees who wish to maintain an effective CAP must submit a license amendment request that includes a description of the proposed CAP. This NUREG provides guidance to the NRC staff on how to determine—from a licensee's CAP license amendment request submittal (CAP submittal)—that the proposed CAP is *acceptable*. After the NRC staff determines that the licensee's CAP is acceptable, the CAP submittal will be incorporated into the license through a license condition, in accordance with the evaluation findings in Section 6 of this NUREG. The NRC staff will then verify implementation of the CAP using a CAP inspection procedure. The purpose of the CAP inspection procedure is to verify that the licensee's acceptable CAP is implemented in accordance with the approved CAP submittal and the licensee's CAP implementing procedures. After the NRC inspection verifies CAP implementation, the NRC will consider the CAP to be *effective*. The conclusion that the licensee's CAP has been determined to be effective will be documented in publicly available inspection reports. Once the licensee's CAP is effective and the criteria in Section 2.3.2a in the NRC Enforcement Policy are met, the NRC staff will start dispositioning Severity Level IV violations as noncited violations. The CAP's implementation will be verified periodically by NRC inspectors in accordance with CAP inspection procedure.

Once the NRC has accepted the licensee's CAP and incorporated it into the license through a license condition, the licensee will be required to meet its CAP commitments.

1. PURPOSE OF REVIEW

The purpose of the review is to determine whether the licensee's corrective action program (CAP) is adequate to support the safe operation of the facility and to identify and correct conditions adverse to safety and security.

This guidance expands on existing guidance in NUREG-1520, "Standard Review Plan for the Review of a License Application for a Fuel Cycle Facility," and NUREG-1962, "Guidance on the Implementation of Integrated Safety Analysis Requirements for 10 CFR Part 40 Facilities Authorized to Possess 2,000 Kilograms or More of Uranium Hexafluoride—Draft Report for Comment," in relation to what the U.S. Nuclear Regulatory (NRC) staff considers an acceptable CAP for fuel facilities.

Title 10 of the *Code of Federal Regulations* (10 CFR) 70.4, "Definitions," defines management measures as including other quality assurance (QA) elements. Section 11.4.3.8, "Other Quality Assurance Elements," of NUREG-1520 states that other QA elements may include some or all of the following elements:

- organization
- QA program
- design control
- procurement document control
- instructions, procedures, and drawing control
- document control
- control of purchased items
- identification and control of items
- control of processes
- inspection
- test control
- control of measuring and test equipment
- handling, storage, and shipping
- inspection, test, and operating status
- control of nonconforming items
- corrective action
- QA records
- audits

For corrective action, Section 11.4.3.8 of NUREG-1520 states (see pages 11-19), "the applicant [or licensee] should specify provisions for promptly identifying conditions adverse to quality and correcting them as soon as practicable."

2. RESPONSIBILITY FOR REVIEW

<u>Primary</u>:   NRC quality assurance reviewer

<u>Secondary</u>:   Licensing project manager

<u>Supporting</u>:   Regional project inspector

3. AREAS OF REVIEW

The licensee's CAP submittal should include the proposed license amendment and supporting program information necessary to address the acceptance criteria specified below.

The specific areas of review of a licensee's CAP submittal are as follows:

- policies, programs, and procedures
- identification, reporting, and documentation of safety and security issues
- significance classification and causal evaluation of safety and security issues
- development and implementation of corrective actions
- assessment of corrective action and program effectiveness

4. ACCEPTANCE CRITERIA

The CAP should be determined acceptable if:

(1)   CAP policies and procedures are established and described indicating terminology definitions, CAP expectations, requirements, personnel responsibilities, and implementation processes.

The CAP organization is described and includes an independent reviewing organization[1] that is auditable and independent of the licensee's production organization. Facility management commits to provide the independent reviewing organization sufficient authority, access to work areas, and organizational independence to perform its responsibilities. The independent reviewing organization reviews and documents concurrence with CAP policies and procedures and revisions thereto.

Specific responsibilities within the CAP may be delegated, but the licensee retains the responsibility for the program's effectiveness and for performing periodic audits and assessments. The CAP requires that delegation authority is documented in writing.

(2)   The CAP includes prompt identification, documentation, and reporting of safety and security issues (i.e., conditions adverse to safety or security). The CAP requires all personnel to identify conditions adverse to safety or security.

(3)   Criteria for classifying the significance of conditions adverse to safety or security (i.e., significant or non-significant) are established. Conditions adverse to safety or security include failures, malfunctions, deficiencies, defective items, out-of-control processes, and nonconformances. For significant conditions adverse to safety or security, the root and contributing causes are determined, the extent of condition and cause are evaluated, and preventive actions are taken to preclude recurrence. Significant conditions adverse to safety or security are those that, if left uncorrected, could have a serious effect on safety or operability.

(4)   Corrective action is promptly developed and initiated following the identification of a condition adverse to safety or security. Conditions and trends that are adverse to safety or security are reported to the appropriate level of management. Follow-up action is taken by the independent reviewing organization, where appropriate, to verify proper implementation of the corrective action. A graded approach is taken to verify proper implementation and close out the corrective actions in a timeframe consistent with the safety or security significance of the issue, with the independent organization reviewing the corrective actions for significant conditions adverse to safety or security.

---

[1]   The independent reviewing organization may be a separate, independent division of the licensee's organization, such as a quality assurance or quality control organization. However, it is also acceptable for the licensee to assign independent review duties to an existing part of the licensee's organization, such as Environmental Health and Safety, provided that the licensee describes this designation in its CAP and commits to ensure that the organization and/or individuals are sufficiently independent, trained, and able to meet the guidance established in this NUREG.

(5)   The effectiveness of the CAP is evaluated by the licensee at regular, specified intervals. Reports of conditions that are adverse to safety or security are analyzed to identify adverse trends in performance. The licensee reviews the ability of the CAP to identify conditions adverse to safety and security, identify NRC reportable events, evaluate the significance, correct the condition, notify management, and prevent recurrence.

5. REVIEW PROCEDURES

For each review area specified in Section 3, the review procedure is provided below. These review procedures are based on the identified acceptance criteria in Section 4. For deviations from these specific acceptance criteria, the NRC staff should review the licensee's evaluation of how the proposed alternatives to the acceptance criteria provide an adequate method to determine that the CAP submittal is acceptable. Figure 1 illustrates a generic corrective action process that can be used for reference by licensees in the development and implementation of their CAPs.

5.1   Policies, Programs, and Procedures

The NRC reviewer should confirm that the licensee describes the CAP expectations and requirements. The reviewer should confirm that the licensee commits to implementing processes in policies, programs, and/or procedures that apply to and are implemented across the licensee's organization and licensed operations.

The reviewer should verify that the licensee describes terminology that will be used to implement its CAP. Terminology should include, as a minimum, conditions adverse to safety or security and significant conditions adverse to safety or security (or equivalent terminology designated by the licensee).

The reviewer should confirm that the licensee has described the independent reviewing organization, including identification of the organization, its responsibilities, and the commitment to provide the independent reviewing organization sufficient authority, access to work areas, and organizational independence to perform its responsibilities. In the case that the reviewing organization has concurrent duties, the reviewer should confirm that the licensee has described how the organization will address a possible conflict of interest.

The reviewer should ensure that personnel responsibilities are defined in sufficient detail to ensure effective processing of conditions adverse to safety and security. The reviewer should also verify that the CAP submittal describes the use of delegation authority. The licensee may delegate specific responsibilities within the CAP in writing; however, the licensee should retain responsibility for ensuring the program's effectiveness and for performing periodic audits and assessments.

5.2   Identification, Reporting, and Documentation of Safety and Security Issues

The reviewer should verify that the CAP submittal requires the licensee contractors, staff, supervisors, and managers to report safety and security issues in a manner that supports the timely and effective assessment and correction of the issues. The reviewer should verify that the CAP submittal describes training requirements for employees to ensure that they are able to identify adverse conditions and understand their CAP responsibilities.

The reviewer should verify that the CAP submittal requires the documentation of safety and security issues from identification to closeout. The reviewer should verify that the CAP describes the licensee's process for tracking and trending of issues and for reporting to the NRC, as needed.

5.3 Significance Assessment and Causal Evaluation of Safety and Security Issues

The reviewer should verify that the licensee's CAP submittal contains a process for evaluating the actual and potential significance of issues. The licensee's assessment should enable the organization to appropriately apply a graded risk approach, based on the issue's significance, to the timing and scope of response to the issues, including the depth and detail of the causal evaluation. For significant conditions adverse to safety or security, the licensee's application of its causal evaluation process routinely enables it to adequately identify the issue's root cause and the contributing factors.

For example, criteria for assessing the significance of conditions adverse to safety and security may include the following:

- impact on health and safety of workers, the public, and environment

- importance in meeting regulatory requirements

- impact on reliability, availability, or maintainability of the equipment important to nuclear safety or security at the facility

- consequence of recurrence or likelihood of recurrence if not corrected

- potential to impact other items or activities beyond the specific occurrence where it may have greater impact

Significant conditions adverse to safety or security may include the following:

- trend of multiple conditions adverse to safety or security

- deficiencies in design, manufacturing, construction, testing, or process requiring substantial rework, repair, or replacement

- damage to a structure, system, component, or facility requiring substantial repairs

- a non-conservative error detected in a computer program or design input after it has been implemented or released for use;

- repeated failure to implement a portion of an approved procedure

Figure 1.- Corrective Action Flow Chart

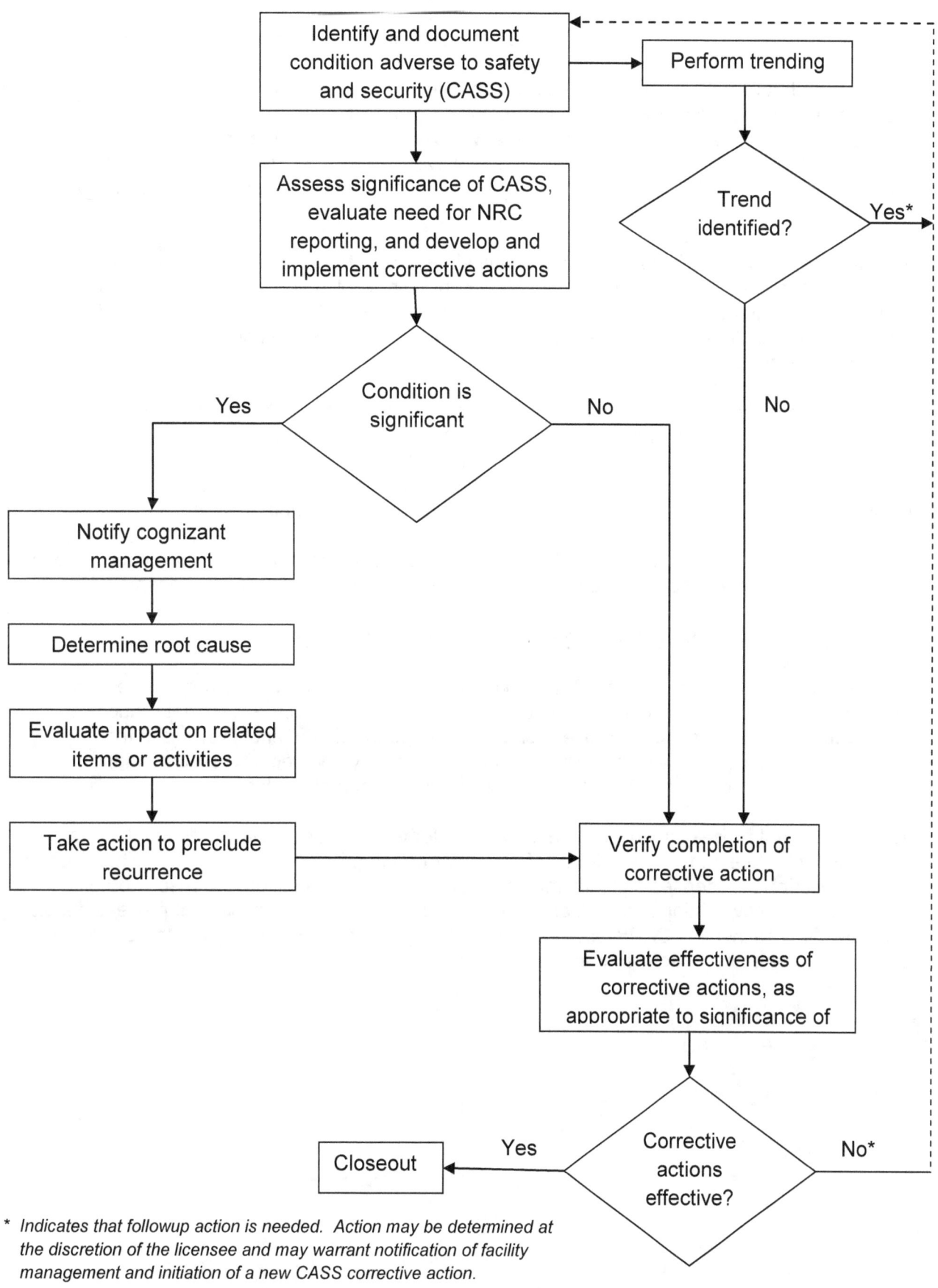

* *Indicates that followup action is needed. Action may be determined at the discretion of the licensee and may warrant notification of facility management and initiation of a new CASS corrective action.*

13

5.4   Development and Implementation of Corrective Actions

The reviewer should verify that the CAP submittal describes the licensee's process for the development and implementation of corrective actions for safety and security issues. The description should include measures that will be taken by the licensee to prevent recurrence of the same issue or the occurrence of similar significant conditions adverse to safety or security. The licensee should also describe actions that will be taken to verify the completion and proper implementation of corrective actions, including involvement by the independent reviewing organization (or other designated party or position for non-significant safety and security issues).

The reviewer should verify that the CAP submittal describes the process by which the licensee will ensure timeliness of corrective action implementation, verification, and close out in a manner commensurate with the safety or security significance of the issues identified. The process should include timeliness expectations in order to ensure prompt resolution of safety and security issues.

For significant conditions adverse to safety or security, the reviewer should verify that the CAP submittal includes measures that the licensee will take to evaluate the extent to which other items and activities, including work in progress, may be affected so that appropriate action can be taken. The reviewing organization should also review the corrective actions developed for significant conditions adverse to safety and security prior to their implementation.

The reviewer should verify that the CAP submittal contains the commitment to report conditions and trends that are adverse to safety or security to the appropriate level of management.

5.5   Assessment of Corrective Action and Program Effectiveness

The reviewer should verify that the CAP submittal describes a CAP assessment process that enables the organization to identify and correct program performance issues that reduce CAP effectiveness. The assessment process should evaluate the CAP's effectiveness in the identification, reporting, assessment, and correction of safety and security issues and the prevention of the recurrence of the same issues or occurrence of similar issues.

The assessment process should include measures for reviewing conditions adverse to safety and security to determine the existence of adverse trends and repetitive problems. The process should describe the licensee's commitments for ensuring the recognition and resolution of ineffective corrective actions, trends, and performance issues. The assessment process should evaluate the timeliness of CAP elements and the ability of the CAP to follow up and enable close out of corrective actions that are past their due date.

6. EVALUATION FINDINGS

The NRC evaluation should verify that the information provided in the licensee's CAP submittal satisfies the acceptance criteria in Section 4. Based on this evaluation, the staff should conclude that the licensee's CAP is acceptable. The reviewer should write a safety evaluation report that includes a summary statement of what was evaluated and the basis for the reviewer's conclusions that the acceptance criteria have been satisfied and the CAP is acceptable.

7. REFERENCES

American National Standards Institute/American Society of Mechanical Engineers Standard, "Quality Assurance Requirements for Nuclear Facility Applications," ANSI/ASME NQA-1-2008.

Code of Federal Regulations, Chapter I, Title 10, "Energy," Part 70, "Domestic Licensing of Special Nuclear Material."

Code of Federal Regulations, Chapter I, Title 10, "Energy," Part 50, "Domestic Licensing of Production and Utilization Facilities."

Code of Federal Regulations, Chapter I, Title 10, "Energy," Part 40, "Domestic Licensing of Source Material."

Code of Federal Regulations, Chapter I, Title 10, "Energy," Part 21, "Reporting of Defects and Noncompliance."

U.S. Nuclear Regulatory Commission, Chapter 11, "Management Measures," NUREG-1520 Revision 1, "Standard Review Plan for the Review of a License Application for a Fuel Cycle Facility," May 2010. Agencywide Documents Access and Management System (ADAMS) Accession No. ML101390110.

U.S. Nuclear Regulatory Commission, Information Notice 96-28, "Suggested Guidance Relating to Development and Implementation of Corrective Action," May 1996. ADAMS Accession No. ML003726705.

U.S. Nuclear Regulatory Commission, Section 17.1, "Quality Assurance during the Design and Construction Phases," Revision 2, NUREG-0800, "Standard Review Plan for the Review of Safety Analysis Reports for Nuclear Power Plants: LWR Edition," July 1981. ADAMS Accession No. ML052350349.

U.S. Nuclear Regulatory Commission, Section 17.2, "Quality Assurance during the Operations Phase," Revision 2, NUREG-0800, "Standard Review Plan for the Review of Safety Analysis Reports for Nuclear Power Plants: LWR Edition," July 1981. ADAMS Accession No. ML052350361.

U.S. Nuclear Regulatory Commission, Section 17.3, "Quality Assurance Program Description," Revision 0, NUREG-0800, "Standard Review Plan for the Review of Safety Analysis Reports for Nuclear Power Plants: LWR Edition," August 1990. ADAMS Accession No. ML052350376.

U.S. Nuclear Regulatory Commission, Section 17.5, "Quality Assurance Program Description—Design Certification, Early Site Permit and New License Applicants," Initial Issuance, NUREG-0800, "Standard Review Plan for the Review of Safety Analysis Reports for Nuclear Power Plants: LWR Edition," March 2007. ADAMS Accession No. ML063190019.

U.S. Code of Federal Regulations, Chapter I, Title 10, "Energy," Part 70, "Domestic Licensing of Special Nuclear Material."

U.S. Code of Federal Regulations, Chapter I, Title 10, "Energy," Part 50, "Domestic Licensing of Production and Utilization Facilities."

U.S. Code of Federal Regulations, Chapter I, Title 10, "Energy," Part 40, "Domestic Licensing of Source Material."

U.S. Code of Federal Regulations, Chapter I, Title 10, "Energy," Part 21, "Reporting of Defects and Noncompliance."

U.S. Nuclear Regulatory Commission, "Management Measures," Chapter 11 NUREG-1520 Revision 1, "Standard Review Plan for the Review of a License Application for a Fuel Cycle Facility," May 2010. ADAMS Accession No. ML101390110.

U.S. Nuclear Regulatory Commission, "Suggested Guidance Relating to Development and Implementation of Corrective Action," Information Notice 96-28, May 1996. ADAMS Accession No. ML003726705.

U.S. Nuclear Regulatory Commission, "Quality Assurance During the Design and Construction Phases," Section 17.1 Revision 2 NUREG-0800, "Standard Review Plan for the Review of Safety Analysis Reports for Nuclear Power Plants: LWR Edition," July 1981. ADAMS Accession No. ML052350349.

U.S. Nuclear Regulatory Commission, "Quality Assurance During the Operations Phase," Section 17.2 Revision 2 NUREG-0800, "Standard Review Plan for the Review of Safety Analysis Reports for Nuclear Power Plants: LWR Edition," July 1981. ADAMS Accession No. ML052350361.

U.S. Nuclear Regulatory Commission, "Quality Assurance Program Description," Section 17.3, Revision 0, NUREG-0800, "Standard Review Plan for the Review of Safety Analysis Reports for Nuclear Power Plants: LWR Edition," August 1990. ADAMS Accession No. ML052350376.

U.S. Nuclear Regulatory Commission, "Quality Assurance Program Description – Design Certification, Early Site Permit and New License Applicants," Section 17.5 Initial Issuance NUREG-0800, "Standard Review Plan for the Review of Safety Analysis Reports for Nuclear Power Plants: LWR Edition," March 2007. ADAMS Accession No. ML063190019.

NRC FORM 335
(12-2010)
NRCMD 3.7

U.S. NUCLEAR REGULATORY COMMISSION

BIBLIOGRAPHIC DATA SHEET

(See instructions on the reverse)

1. REPORT NUMBER
(Assigned by NRC, Add Vol., Supp., Rev., and Addendum Numbers, If any.)

NUREG-2154
DRAFT

2. TITLE AND SUBTITLE

Acceptability of Corrective Action Programs for Fuel Cycle Facilities
Draft for Comment

3. DATE REPORT PUBLISHED

MONTH	YEAR
January	2013

4. FIN OR GRANT NUMBER

5. AUTHOR(S)

Sabrina Atack

6. TYPE OF REPORT

Technical

7. PERIOD COVERED (Inclusive Dates)

8. PERFORMING ORGANIZATION - NAME AND ADDRESS (If NRC, provide Division, Office or Region, U. S. Nuclear Regulatory Commission, and mailing address; if contractor, provide name and mailing address.)

Division of Fuel Cycle Safety and Safeguards
Office of Nuclear Material Safety and Safeguards
US Nuclear Regulatory Commission
Washington, DC 20555-0001

9. SPONSORING ORGANIZATION - NAME AND ADDRESS (If NRC, type "Same as above"; if contractor, provide NRC Division, Office or Region, U. S. Nuclear Regulatory Commission, and mailing address.)

Same as Above

10. SUPPLEMENTARY NOTES

11. ABSTRACT (200 words or less)

The NRC staff has revised Section 2.3.2 of the NRC Enforcement Policy (ADAMS Accession No. ML12340A295) to disposition Severity Level IV violations for Fuel Cycle Facilities as non-cited violations if the NRC determines that the licensee's CAP is effective, the licensee enters the violation in its CAP, and other criteria in Section 2.3.2 of the Enforcement Policy are met. The purpose of this draft NUREG, "Acceptability of Corrective Action Programs for Fuel Cycle Facilities," is to provide guidance to the NRC staff on how to determine, from a licensee's CAP licensing submittal, that a CAP is acceptable. After the NRC staff determines that the CAP is acceptable, the CAP licensing submittal will be incorporated into the license and implementation of the CAP will be verified by an NRC inspection using a CAP inspection procedure. After the NRC inspection verifies that the licensee has implemented its CAP in accordance with the license and the licensee's CAP implementing procedures, then the NRC will consider the CAP to be effective for the purposes of Section 2.3.2 of the Enforcement Policy.

12. KEY WORDS/DESCRIPTORS (List words or phrases that will assist researchers in locating the report.)

QA, CAP, quality assurance, corrective action, corrective action program, NCV, non-cited violation, noncited violation,

13. AVAILABILITY STATEMENT

unlimited

14. SECURITY CLASSIFICATION

(This Page)

unclassified

(This Report)

unclassified

15. NUMBER OF PAGES

16. PRICE

NUREG-2154
Draft

Acceptability of Corrective Action Programs for Fuel Cycle Facilities

January 2013